Polymers Among Us

by **Conrad J. Storad** illustrated by **Jeff Yesh**

STORY MONSTERS PRESS
Everything Children's Books
An imprint of Story Monsters® LLC
Chandler, Arizona, USA

First edition published September 2026

Linda F. Radke, Publisher
Story Monsters® Press
An imprint of Story Monsters® LLC
4696 W. Tyson Street
Chandler, AZ 85226
480-940-8182
Publisher@storymonsters.com
www.StoryMonstersPress.com

Publisher's Cataloging-in-Publication
(Provided by Cassidy Cataloguing Services, Inc.).
Names: Storad, Conrad J., author. | Yesh, Jeff, 1971- illustrator.
Title: Polymers among us / by Conrad J. Storad ; illustrated by Jeff Yesh.
Description: Chandler, Arizona, USA : Story Monsters® Press, an imprint of Story Monsters LLC, [2026] | Audience: Juvenile. | Summary: What the heck is a polymer? What do sneakers, toothbrushes, toys, and even space technology have in common? They're all made with polymers! Join the curious Story Monster and his fun-loving polymer guides, Pauly and Penny, on a colorful comic-style adventure that makes polymer science fun, easy to understand, and full of surprises. Bursting with fascinating facts, real-world connections, and hands-on discoveries, Polymers Among Us sparks curiosity, inspires big imaginations, and reveals the amazing science hiding in everyday life--while encouraging tomorrow's scientists, engineers, and innovators.--Publisher.
Identifiers: ISBN: 9781589854048 (paperback) | 9781589854055 (hardcover) | 9781589854062 (ebook) | 9781589854079 (audiobook)
Subjects: LCSH: Polymers--Juvenile literature. | Science--Vocational guidance--Juvenile literature. | Engineering--Vocational guidance--Juvenile literature. | CYAC: Polymers. | Science. | Engineering.
Classification: LCC: QD381.8 .S76 2026 | DDC: 668.9--dc23

ISBN: (paperback) 978-1-58985-404-8
ISBN: (hardcover) 978-1-58985-405-5
ISBN: (eBook) 978-1-58985-406-2
ISBN: (audiobook) 978-1-58985-407-9; Story Monsters® Little Read-Alongs™

Creative Story Strategist: Paula Hurd
Illustrations and page design: Jeff Yesh

Printed in the United States of America

To young readers living in Akron,
Barberton, and throughout Northeast Ohio.
Never stop reading. Never stop learning!

– CJS

AUTHOR'S FOREWORD

Let's get started...

Polymer.

Say the word this way: **paa-luh-mer**

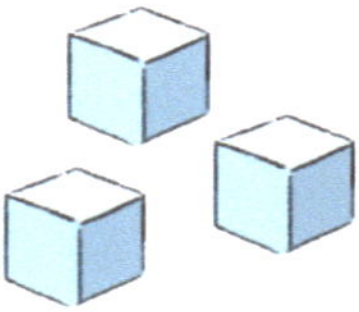

The word polymer means "many parts." One part is called a monomer. Polymers are lots of monomers connected together.

Think about building blocks. All the blocks look the same. You can link them to make a long chain. Stack them and make a flat wall. You can build lots of shapes.

People and animals are made of polymers. Trees are made of polymers. Bouncy balls and lots of toys are made of polymers. So are the tires on cars and trucks and airplanes.

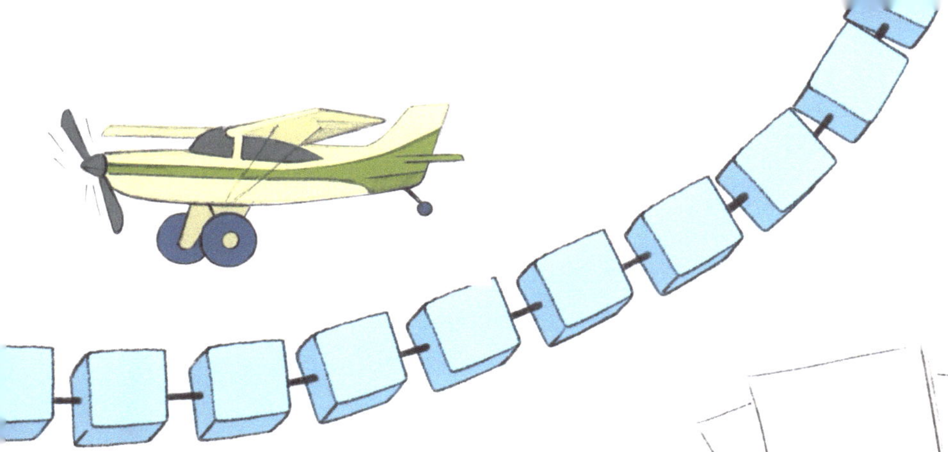

People use polymers every day.

The page you are reading right now is made of paper. Paper is made of cellulose. It is a natural polymer.

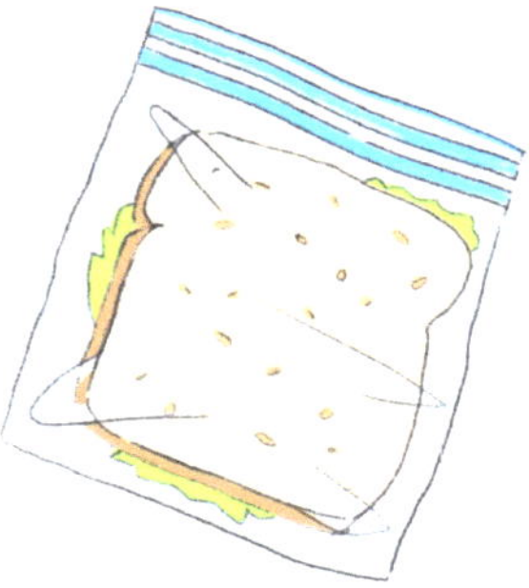

The sandwich in your lunch bag is wrapped in a thin sheet of plastic. The plastic is a polymer. So are parts of the clothes and shoes you wear.

Humans have learned to make all kinds of polymers. These are called synthetic polymers. Rubber and plastics are just a few.

Polymers are everywhere.

Follow the Story Monster as it travels to learn more.

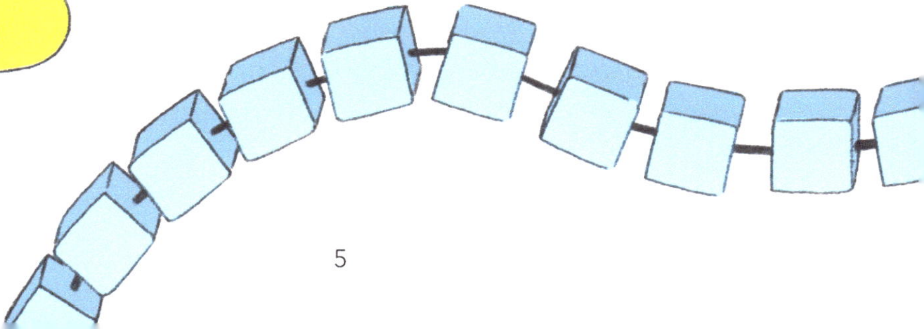

HAVE YOU MET A STORY MONSTER?

Since the earliest days of painting on cave walls, humans have never been alone in their love of stories. Something else has always been watching ... and listening.

These watchers are shy. These listeners don't want to be seen. They go by many names and come in all shapes and sizes. They are Story Monsters.

Story Monsters love stories—and they love to learn. They are also shapeshifters. That means they can stretch, shrink, bounce, twist, and transform into all kinds of shapes. One might be as small as a pea, while another could grow as tall as an oak tree. They might look like big red dogs or curious little monkeys. They could even appear as wooden puppets with long noses ... or as a cat in a tall, striped hat.

Story Monsters are remarkable. But they are not alone. Did you know there are other things that can bend, twist, and change shape into different forms?

Polymers! What are polymers anyway?

One curious Story Monster is about to find out, setting off on an adventure into the squishy, stretchy, surprising world of polymers!

Early one morning, not long ago...

a little monster set out on an adventure.

The little monster wandered into the hushed woods just as the sun painted the sky orange and gold. Sunbeams peeked through the branches, making its lime green skin sparkle. The morning air was cool and crisp, perfect for the start of a new adventure.

The little monster felt excited. Somehow, it knew today would be filled with new discoveries. And learning new things was one of its favorite things to do.

Wind rustled the leaves.
A soft voice whispered...

You need to learn about the Polymers.

Polymers are your cousins.

The monster's thin eye stalks swiveled in every direction, searching for the voice.

The voice seemed to come from the trees themselves.

Polymers are part of many things.

There is much for you to learn.

Go to the Rubber City.
You're very close. It's called Akron, Ohio.
Where was the voice coming from?
That area is home to many families of Polymers.
Suddenly the little monster turned into a question mark.
Just follow the Polymer Pathway.

Rainbow shapes began to appear hovering above the ground. Some were round. Some were square. There were so many colorful shapes.

The shapes formed a sparkly path in front of the little monster.

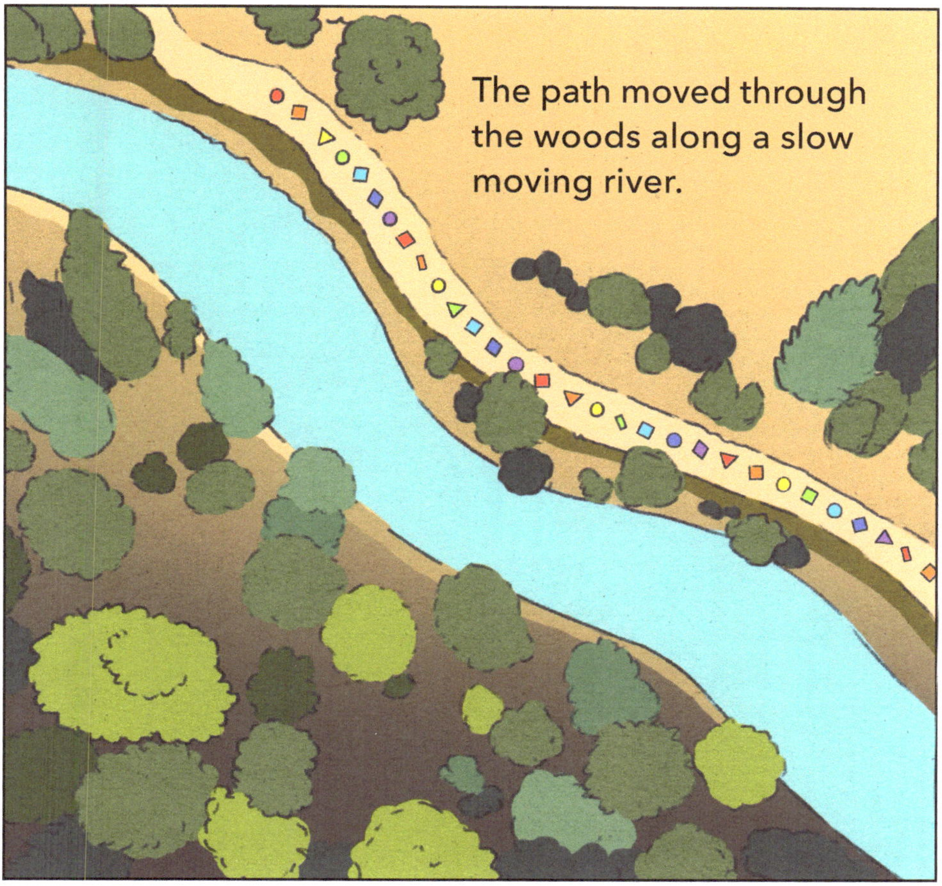

The little monster moved quickly along the path, chasing the shapes.

The shapes disappeared as the little monster moved right through them.

Soon the path changed. It became a wide, concrete trail.

The monster came to a tall group of signs that pointed in lots of directions. They showed the names of many places.

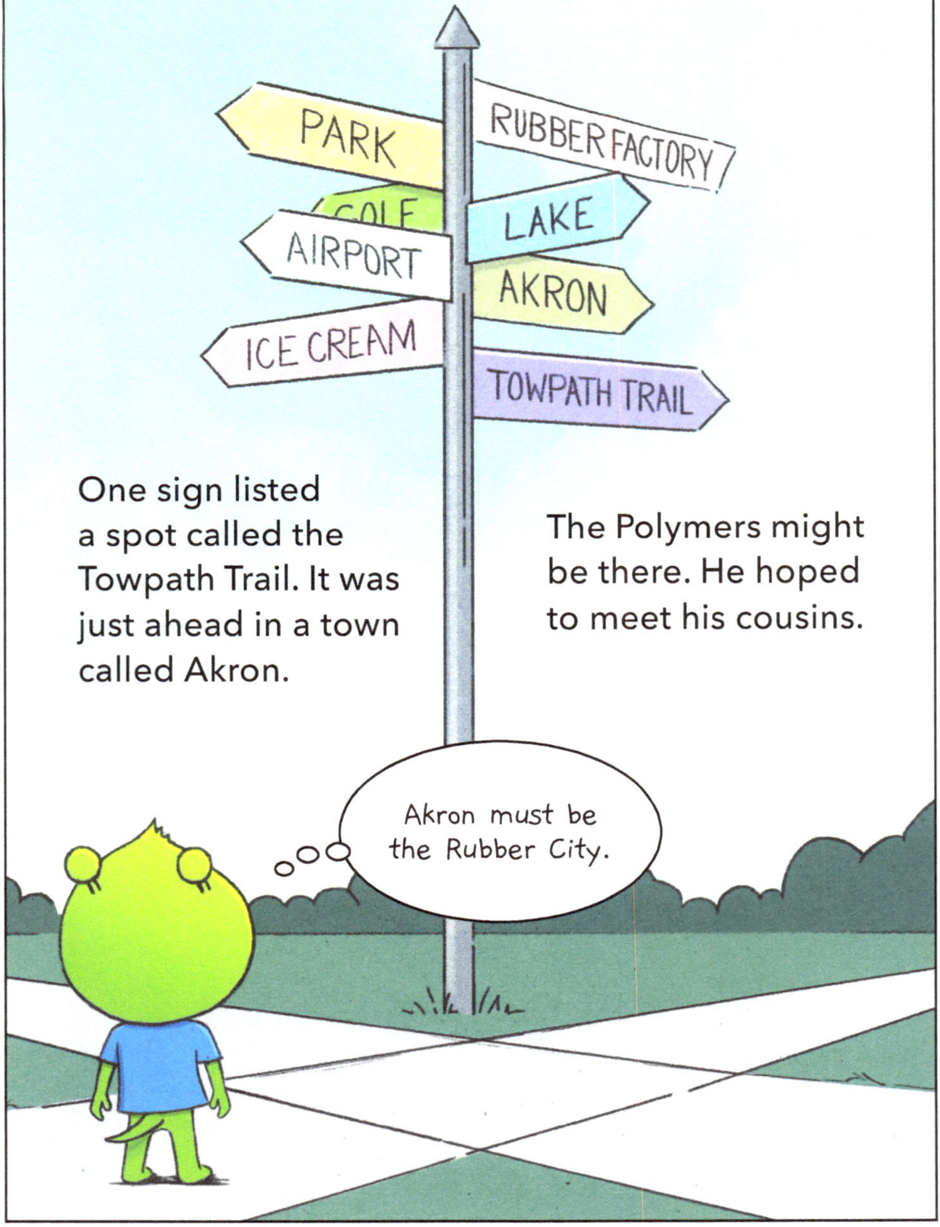

One sign listed a spot called the Towpath Trail. It was just ahead in a town called Akron.

The Polymers might be there. He hoped to meet his cousins.

The monster followed the path until it stood under a wooden archway painted purple and white. Letters on the arch spelled out **OHIO & ERIE CANAL TOWPATH TRAIL.**

Across the street was a long, low building with a big sign out in front.

The little monster used his superior eyesight to zoom in on the words. A few words came into focus.

There was a picture of what looked like a magic lamp. This had to be the place!

A loud voice startled the monster.

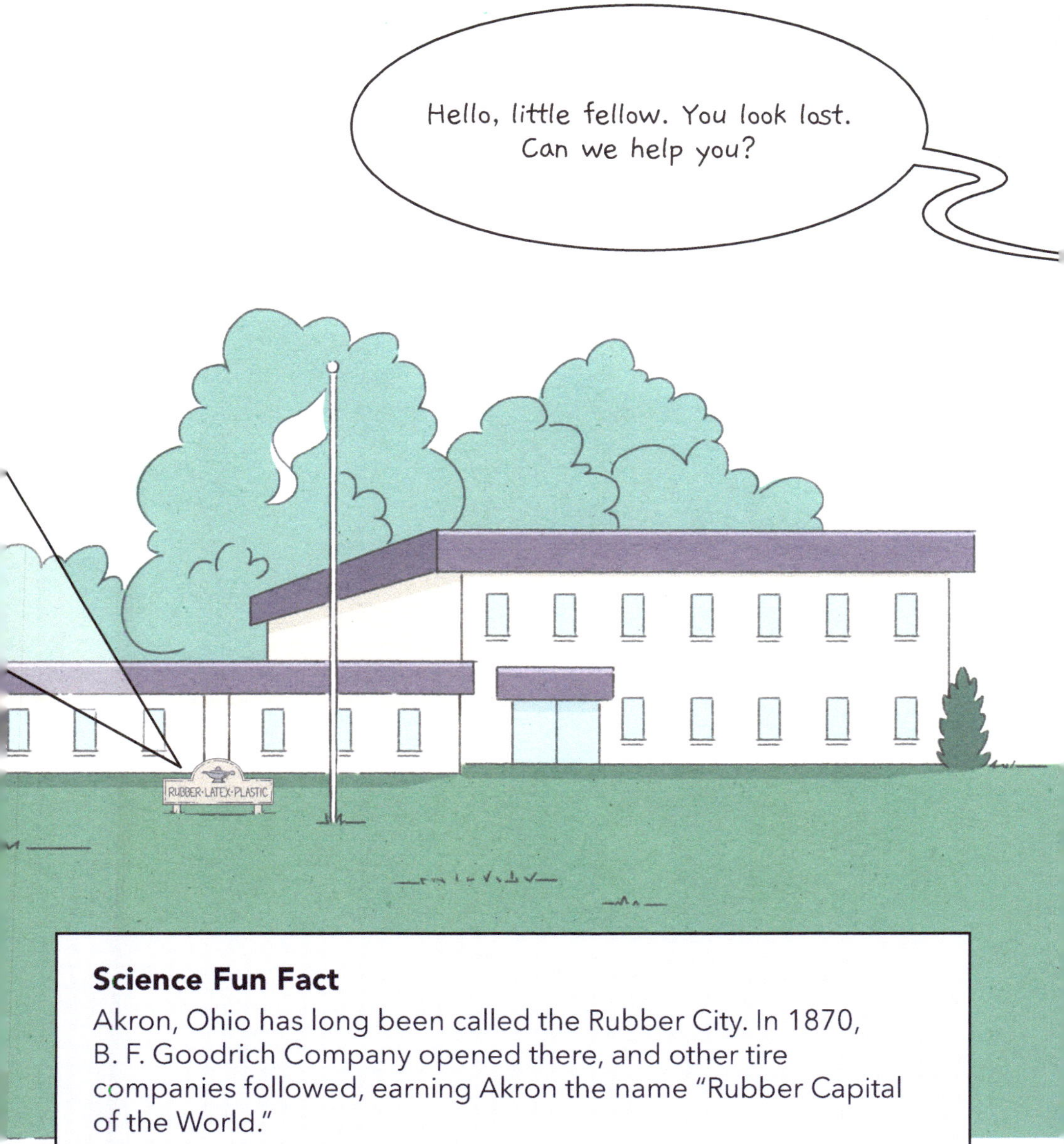

Science Fun Fact

Akron, Ohio has long been called the Rubber City. In 1870, B. F. Goodrich Company opened there, and other tire companies followed, earning Akron the name "Rubber Capital of the World."

Later, the University of Akron began leading the study of polymers–materials like rubber and plastic–and started the nation's first Ph.D. program in Polymer Chemistry. Today, Akron is still a world leader in polymer science.

WELCOME TO THE RUBBER CITY!
Well, almost. You're still a few miles away.

The voice came from a happy, round red blob with one big eye. It seemed to appear out of nowhere.

Science Fun Fact

Polymers are all around the world, even if you don't always see them. Their strength, flexibility, and durability come from their special structure, making them important for creating new products.

The smaller blob, called Penny, seemed to sparkle in the light. She was an ***elastomer***.

She stretched herself longer ... and longer ... and longer!

Then–she snapped right back to her normal shape.

Next she flattened herself like a pancake.

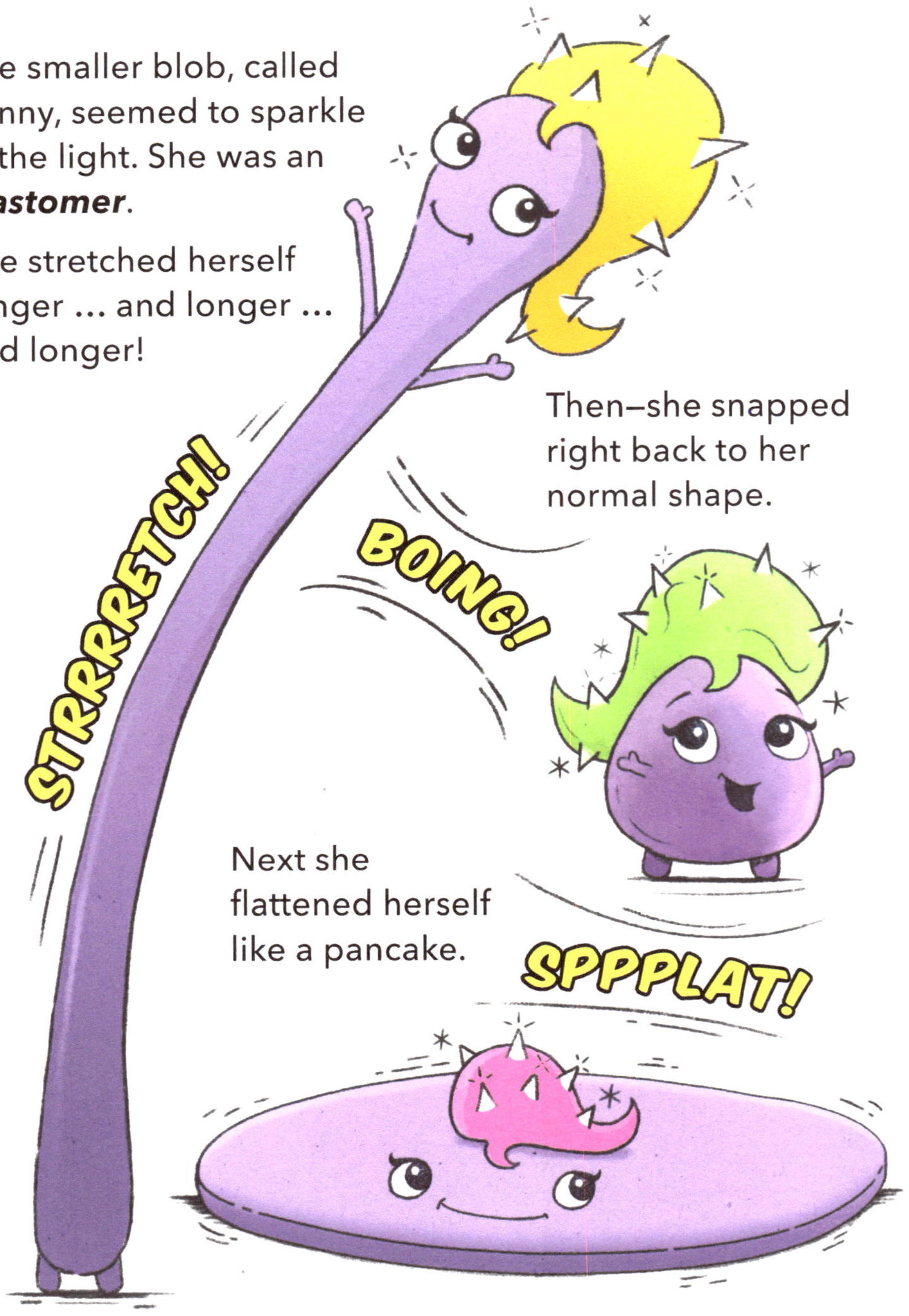

Science Fun Fact

Stretchy polymers are called elastomeric. Many toys are made using polymers. There are stretchy balls that expand and then they "remember" their original shape–and bounce back into a ball!

The monster said nothing.

Then it shimmered with a purplish glow and began to change its shape.

In seconds it looked exactly like Pauly.

And then it looked like Penny.

Science Fun Fact

Shape-shifting polymers can exhibit dramatic transformations–folding, bending, or expanding–making them highly valuable for various applications.

Pauly's big eye blinked. He smiled. Penny's hair turned bright gold. She giggled. Her voice sounded like jingle bells.

The monster turned back into itself ... and smiled. Its two white fangs seemed to glow.

Science Fun Fact

Polyethylene terephthalate (PET) and polyester are the same plastic. PET is used for water bottles and food containers. When made into soft fibers for clothing, it's called polyester.

The monster's four eyes opened wide.

All this gooey melting reminded the little monster of its favorite toy.

In its hands appeared a big tub of slime.

Science Fun Fact

Some polymers are like slime. When they're warm, they get soft and stretchy. But if they get too hot, they fall apart–scientists call that thermal degradation.

The trio found a shady spot under the bridge along the river to cool off.

This was the part of the Towpath Trail the little monster had read about.

Science Fun Fact

Rubber duckies are made of polymers–long chains that stick together! These make them soft, squishy, waterproof, and perfect for floating in the tub.

When Great Granny Ester told us about Story Monsters, she explained that shapeshifters like to learn by imitation. They do this by morphing into what they see.

Pauly's entire face became one huge grin.

Okay, cousin, time for the squishy, bendy, and totally science-tastic truth. Get ready to stretch your imagination!

The little monster glowed with excitement. It sat perfectly still, a good listener ready for a great story.

Science Fun Fact

Shape memory polymers are "smart" plastic materials. They can transform into a temporary shape and then return to their original shape. The change is triggered by an external stimulus such as pressure.

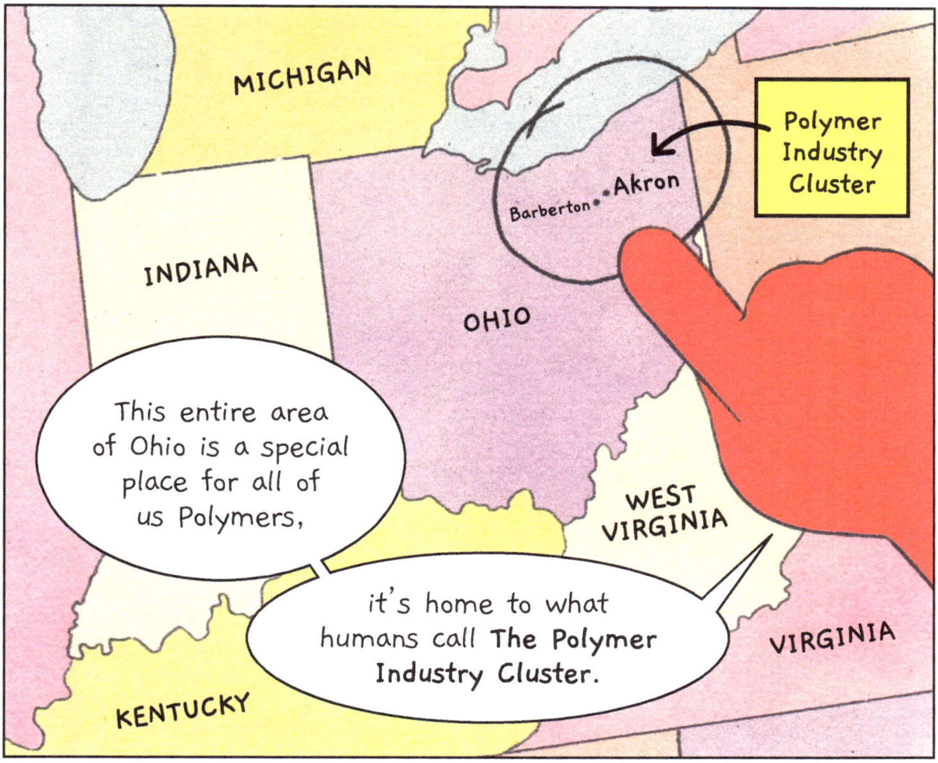

Science Fun Fact

The Polymer Industry Cluster Initiative was launched in 2021. A group called the Elevate Greater Akron Collaborative took a closer look at the region's polymer industry. The idea was to help it reach its full potential.

Pauly flickered.

Science Fun Fact

Animals, plants, and people are made of polymers. Plants use cellulose to build their walls. Our bodies use proteins–like collagen, keratin, and the ones in our muscles–to grow and move. The most famous polymer is DNA.

Science Fun Fact

The stretchy stuff that makes chewing gum chewy is a polymer. Long ago, people got the chewy polymer from tree sap. Today, most chewing gum is made with human-made polymers. Some actually help to make the flavor last longer.

Today, let's talk about Penny and me. And some of our local cousins.
Nylon
We are from a different part of the family. We are synthetic polymers. That means we are made by humans.
Polyester
Polystyrene
Polyvinyl Chloride (PVC)
Polypropylene
Polyethylene
There are lots of us in the Akron area. We have jobs of all kinds.
GAC
GREATER AKRON CHAMBER
THE POLYMER INDUSTRY CLUSTER
The Polymer Industry Cluster has lots of companies in Akron and Barberton and other cities.
The human workers are scientists and engineers. Some are technicians and mechanics. They have all kinds of jobs.

Pauly paused. He rubbed the spot where a chin should be.

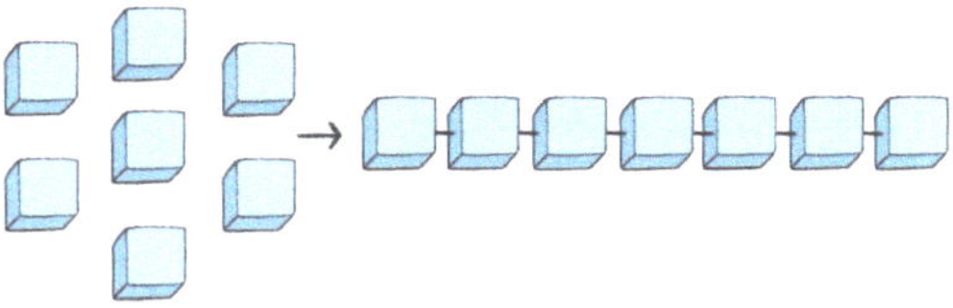

Science Fun Fact

The process through which large chains of monomer molecules combine together to form the polymer is called polymerization.

The little monster flickered. For a few seconds, there was a long string of tiny purple monsters. They all looked the same.

There was a soft buzzing sound.

The little monster looked like a flat sheet.

Then it became a square.

Then a pile of foam.

There was a puff of purple mist. The monster was back.

The little monster changed in a flash.

It formed into a silly pipeline. Parts of it curved left and right and up and down.

Then the little monster changed into colorful action figures. First, it was a dinosaur.

Then it became a superhero. It was having so much fun!

They were gone in a cloud of purple.

Science Fun Fact

Many children's action figures are made from polyvinyl chloride (PVC). The material is durable, versatile, and allows for fine, molded details. PVC is also known for its ability to hold paint well. It also withstands the rough and tumble world of child play.

Just a few more. We have to go soon, it's getting late.
Cousin Ethyl is very famous in the world of plastics. She's big into bottles and windows. All kinds of stuff.
Whoa!
Then he talked about Cousin Kev.
Kev's middle name is Lars. He does work for the military. Kev is very tough. He's into helmets and body armor.
Don't forget Cousin Zippy! He's the wrapper of the family.
He loves food and ... keeping it fresh!
The little monster had a fun time changing into a "zippy rapper."

The little monster was having a wonderful time. But he had gotten quite a workout! After so many shape shifts, it was time to go.

Pauly and Penny stood near the monster. They were back near the Towpath Trail.

Time for you to head to Akron! Akron is known as the Rubber Capital of the World! Scientists and engineers at the University of Akron study polymers of all kinds.
RUBBER CAPITAL OF THE WORLD
AKRON
UA
That's where gazillions of rubber tires for cars and trucks were made. Blimps and other stuff, too.
Rubber is a very important polymer for humans. Especially in this part of Ohio.
Follow the Polymer Pathway for a few more miles. You can visit Akron.
OHIO
N
W E
S
AKRON
POLYMER PATHWAY

The monster thought about the form of rubber he should morph into before traveling to Akron.
A bouncing basketball?
No, too slow.
Maybe a bicycle tire?
No, too small and narrow.
AHA! A huge tractor tire!
Pauly and Penny laughed as they waved goodbye.
VRRROOOMMM!

The little monster remembered that the cool metal tires built for NASA's moon buggy were created in Akron, Ohio.

As it rolled down the Polymer Pathway toward Akron, the little monster morphed one last time. Now it looked like a tire from the Mars Rover named *Curiosity*.

In a puff of sparkling purple mist, the little monster rolled into Akron. **He was very excited!**

What will Story Monster discover on his next adventure?

Science Fun Fact

As part of the Apollo Program, NASA partnered with Akron-based Goodyear Tire & Rubber Company to develop specialized airless "spring tires." The tires featured a network of 800 interwoven springs. They were designed for driving over rugged lunar terrain.

READING GUIDE AND LEARNING ACTIVITIES

WELCOME!

In the book *Polymers Among Us*, you learned many interesting facts through a fun story. Science is all around us. It helps explain how the world works. Maybe you will become a scientist one day! There are many kinds of science to study. Polymers are materials that form many of the building blocks of our world.

PLANT STRUCTURE

ACTIVITY 1: IDENTIFY A PLANT

1. Go outside and find a plant you do not know.
2. Take a close-up picture of one of its leaves.
3. Use Google Lens or another plant-identifying app to find its name.

Plants contain a polymer called **cellulose**. Cellulose helps plants stay strong while still bending in the wind.

ACTIVITY 2: CELERY FIBERS

1. Take a stalk of celery.
2. Slowly bend and break it.

Look closely. You will see string-like fibers running along the celery stalk. These fibers contain cellulose. Cellulose helps the celery stand up straight while also letting it bend.

WHAT IS A MONOMER?

MAKE A PAPER CHAIN

You will need:

- One sheet of 8½ x 11-inch paper
- Scissors
- Glue
- A ruler

Directions:

1. Fold the paper in half the long way.
2. Cut along the fold to make two long strips.
3. Stack the strips together.
4. Use a ruler to mark lines every 1 inch.
5. Cut along the lines. You should have about 22 paper strips.
6. Take one strip and glue the ends together to make a circle.
7. Put a second strip through the first circle.
8. Glue the ends of the second strip together.
9. Keep adding strips to make a paper chain.

Think About It

- Is the chain stiff? ______
- Is it flexible? ______
- Can it stretch, twist, or bend? ______

A polymer is made of many small parts called **monomers**. The monomers join together to make long chains. These chains can be strong, flexible, or stretchy depending on what they are made to do.

POLYMERS IN EVERYDAY LIFE

Look in your closet. Find clothing labels on shirts, socks, jackets, or pants.

Look for these words:

- Polyester
- Rayon
- Cotton

Describe one item of clothing:

____

____

____

Think About It

- How do your socks stay up?
- What makes a T-shirt easy to put on?
- Why do car tires have grooves and ridges?
- Why doesn't outside air come through closed windows and doors?

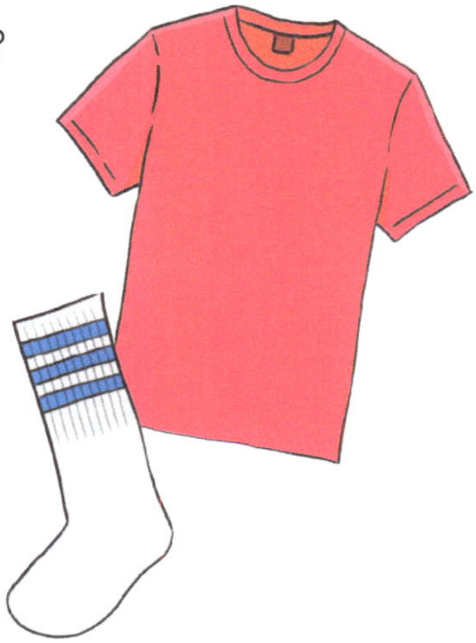

MATCHING ACTIVITY

The prefix **poly-** means "many."

Match each word to the correct category.

Word	**Category**
Polyphonic	______________________
Polymath	______________________
Polyglot	______________________
Polyhedron	______________________
Polynomial	______________________
Polysyllabic	______________________
Polytheism	______________________
Polynesia	______________________
Polyester	______________________

Categories

- Music
- Genius
- Languages
- Geometry
- Math
- Vocabulary
- Religion
- Geography
- Fashion

MAP OF OHIO

1. Put a dot where the City of Akron is located and label it.
2. Find Ohio's capital city. Draw a star and write its name.
3. Name the states that border Ohio.

THE ERIE CANAL

The Erie Canal was an important waterway. Long ago, boats carried goods through canals. Mules often pulled the boats along the canal.

The canal helped businesses move products from one place to another.

Think About It

- Draw the path of the Erie Canal.
- Why is water transportation important for manufacturing and trade?

Here are sources for this activity:

- https://fhwaapps.fhwa.dot.gov/bywaysp/byway/10501/map
- https://geology.com/cities-map/ohio.shtml
- https://www.eriecanal.org/boats-2.html

BE A STORY CREATOR

Choose a polymer object such as:

- A super ball
- Slime
- Modeling clay

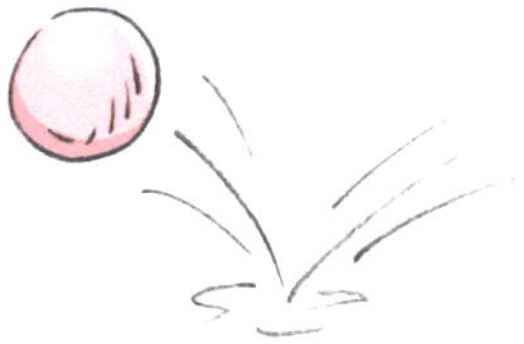

1. Give your object a name.
2. Write a story or poem about its adventure.

Think About It

- Where does it go?
- What happens to it?
- Does it change shape?
- Does it solve a problem?

HANDS-ON ACTIVITY

Make a sign for outside your bedroom door.

Include:

- What room is on the left
- What room is on the right
- What room is straight ahead

Use symbols, arrows, and pictures to make your sign easy to read.

RESEARCH ACTIVITY

Find out what is made in your town or city.

1. Search online using your town's name and the word "manufacturing."
2. Visit a company website.
3. Write down what products the company makes.
4. Determine if each product is made from polymers.

ONLINE POLYMER EXPERIMENTS

Watch online videos about polymers and how they work.

Think About It

- Stretching
- Bending
- Flexibility
- Strength

Write down one new fact you learn from each video.

- **Polymers for Kids!: What Makes Slime Stretch? Science Fun for Elementary Students**
  www.youtube.com/watch?v=IkaSJUEHr0A
- **Kids Experiment: Magic of Polymer**
  www.youtube.com/watch?v=e3A2z4-MbRQ
- **Clayton's Exploration Station: Exploring Polymers!**
  www.youtube.com/watch?v=jrBVnNu_UO8

MATCHING ANSWERS

- Polyphonic – Music
- Polymath – Genius
- Polyglot – Languages
- Polyhedron – Geometry
- Polynomial – Math
- Polysyllabic – Vocabulary
- Polytheism – Religion
- Polynesia – Geography
- Polyester – Fashion

MAZE

Look at the front and back covers. What polymers do you see? Remember, any item may be a polymer, or it might contain one or more types of polymers. Some polymers are on the inside. Some are on the outside.

- The blimp.
- The clouds in the sky. (May contain microplastics from pollution)
- Airplane tires.
- Airplane windows may be plastic.
- Rope and banner behind airplane might be nylon.
- Kite and kite string.
- Little girls and boy. Human bodies contain many types of polymers.
- Picnic cooler, cups, and food containers.
- Picnic table.
- Skateboard and wheels.
- Playdough and slime.
- Whiffle ball bat and ball.
- Kickball.
- Bicycle tires, seat, and bicycle helmet.
- Every blade of grass.
- Plastic sliding board.
- Wood in sliding board tower.
- Balloon and string.
- Basketball.
- Soccer ball.
- Tire swing and rope.
- Trees and leaves.
- Baseball cap.
- Pauly Polymer.
- Penny Polymer.
- The Story Monster's sneakers, socks, and shirt.
- What about the Story Monster itself? You decide.

How many more polymers can you find? They're everywhere!

ACKNOWLEDGEMENTS

This book is the result of a shared commitment to helping young people discover the exciting world of polymers and the opportunities that exist right here in our community.

We extend our sincere gratitude to the members of the Polymer Industry Cluster, whose leadership and dedication help inspire the next generation of innovators, problem-solvers, and makers. Special thanks to the Greater Akron Chamber and ConxusNEO for connecting education, workforce development, and industry, and to Akron Rubber Development Lab for sharing its expertise and helping bring the science of polymers to life.

We are especially grateful to the Barberton Community Foundation for investing in educational opportunities that inspire curiosity and exploration, and to Barberton City Schools for embracing innovative programs that help students discover the exciting careers available in the polymer sector.

Thank you to our educators, employers, community partners, families, and students whose enthusiasm and support made this book possible. Your commitment to learning, discovery, and opportunity continues to strengthen our region and its future.

Together, we are helping ensure that the Polymer Capital of the World is not just part of our history–it is part of their future.

ABOUT THE AUTHOR

Conrad J. Storad is the award-winning author and editor of more than 60 science and nature books for young readers. He also writes a monthly science/nature column called "Conrad's Classroom" in Story Monsters Ink®, a national award-winning magazine. Many of Storad's books reflect his interest and fascination with nature's amazing plants and creatures. An accomplished storyteller, Storad has "edu-tained" more than 1 million students and teachers across the nation with his programs and writing workshops that promote reading and science literacy. After 33 years living and working in Tempe, Arizona, Conrad and wife Laurie now live in their hometown of Barberton, Ohio, where he was a 2017 honoree on the city's Walk of Fame.

ABOUT THE ILLUSTRATOR

Jeff Yesh is a graphic designer and illustrator whose award-winning work has been featured in multiple children's books. He is also the artistic talent behind the Story Monster character and is the graphic designer for the award-winning Story Monsters Ink®, the literary resource for parents, teachers, and librarians. Yesh graduated from Indiana State University with a Bachelor of Fine Arts in Graphic Design. He lives in Indiana with his wife and two daughters.

ABOUT THE PUBLISHER

Story Monsters® Press, an imprint of Story Monsters® LLC, is a publisher of children's books that inspire hope, celebrate differences, and help build character. Select titles may include curriculum guides, along with bonus materials such as coloring pages, puzzle pages, and learning activities designed to complement the stories and support parents and educators in engaging young readers.

Explore more amazing books published by Story Monsters® Press at storymonsters.com/books. To find out how to invite our incredible authors to your school or events, please contact info@storymonsters.com.

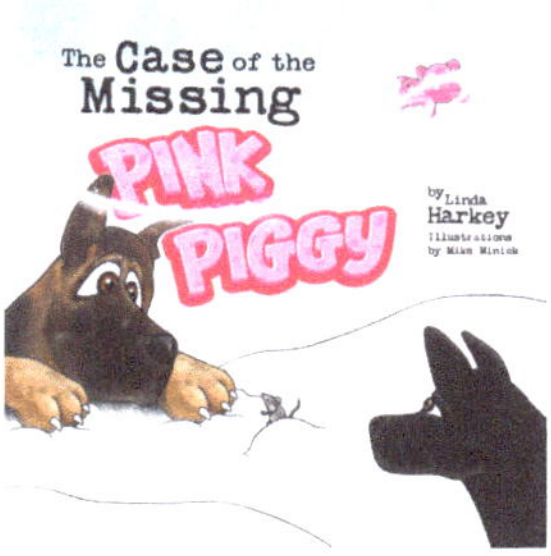

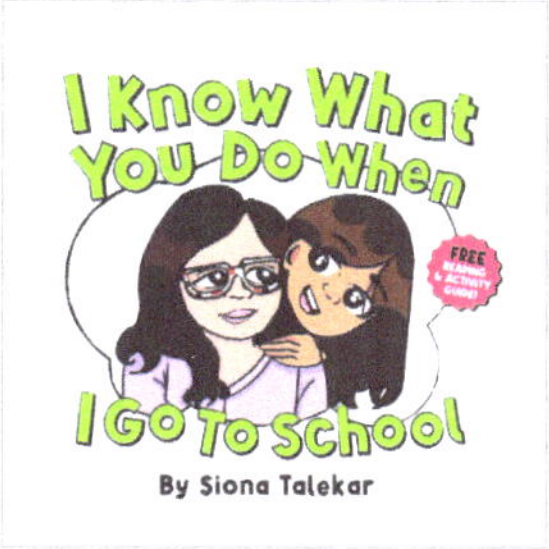

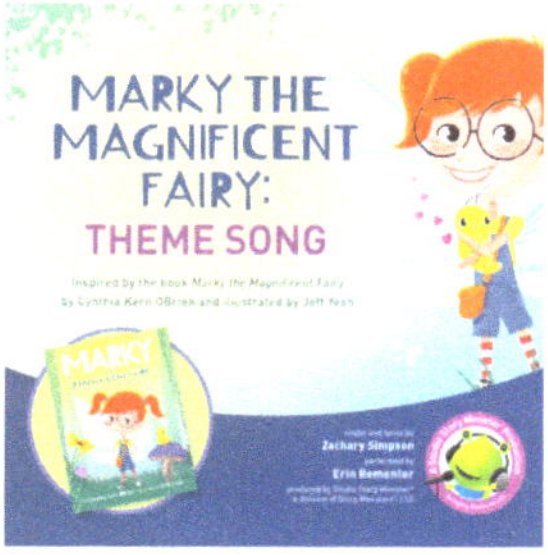

www.ingramcontent.com/pod-product-compliance
Lightning Source LLC
LaVergne TN
LVHW071909071126
851857LV00033B/8